R.O.S. "Return on Safety":

Ensuring a Positive Return on Your Investment in Workplace Safety

By Doug Crann

Here's What's Inside...

5 **Introduction**

6 **Dedication**

6 **Acknowledgements**

7 **R.O.S. "Return on Safety"!**

7 **Companies Who Invest in Workplace Safety Have Lower on the Job Injuries...**

8 **Why Shortcuts in Workplace Safety Always Backfire...**

12 **Why Don't Companies Embrace Workplace Safety?**

15 **Focus on Sustainable Safety, Not Compliance...**

18 **Workplace Safety Leads to Higher Profits and Increased Morale...**

21 **Workplace Safety Is the True Measure of an Organization's Success...**

23 **Four Common Workplace Safety Misconceptions...**

27 **How the Real Test of Your Safety Program Isn't Injuries but Rather: Zero Unsafe Behaviors and Conditions...**

28   **Here's What a Workplace Safety Campaign Consists of...**

32   **Workplace Safety Works for Every Company in Every Industry...**

34   **Here's Exactly How to Get Your Workplace Safety Program Going...**

38   **Here's How to Take Control of Your Safety Responsibilities and Get Your Workplace Safety Program Compliant and Documented...**

Introduction

March 2014
Newmarket, Ontario

If you are struggling with compliance because past attempts at safety improvements were ineffective, unsustainable, counterproductive or too costly, this book will help you understand why many businesses fail at implementing a successful safety program and you will learn an approach that ensures your next investment in safety program improvements will give your business a positive return.

What follows is an interview I had where I discuss what this can look like for your organization.

Enjoy the book!

I hope it changes the way you think about workplace safety improvements and you start seeing workplace safety not as a burden but as an asset which returns 10x the investment.

Always Be Safe,

Doug Crann

Doug Crann, Founder and President
WORKPLACE SAFETY REVOLUTION

Dedication

This book is dedicated to my wife and our 6 amazing children: Marianne, Amie, Nicholas, Dylan, Marissa, Alycia and Amanda, without your hearts and minds to share in this journey, I fear I would not have reached so high.

Acknowledgements

Glenn Mcqueenie, CEO of Keller Williams Referred Realty; Keller Williams Referred Urban, Referred Advance, and Touch 33 Marketing Ltd

Brian Hagen, President and Executive Producer MOS Productions

Marianne Crann, Wife and Owner HR Revolution

R.O.S. "Return on Safety"!

Susan: Hi. Good afternoon. This is Susan Austin and with me today is Doug Crann from Newmarket, Ontario. Welcome Doug.

Doug: Hi, Susan. Great to be speaking with you.

Companies Who Invest in Workplace Safety Have Lower on the Job Injuries...

Susan: Today, we're going to be talking about your book *Return on Safety: Ensuring a Positive Return on Your Investment in Workplace Safety*. Let's just jump right in Doug. Companies that invest in workplace safety, do you find that they have a lower injury rate than companies who don't?

Doug: Not many people will argue that investing in workplace safety has the potential to reduce workplace injuries, but let's be clear; results vary to say the least. Harm reduction is synonymous with investment in workplace safety whether it's training, policies or hiring a full-time Safety Manager. They all have an impact on the likelihood of whether people are getting hurt at work. There's no one who will dispute or argue that fact.

However, despite improving technology and an increase in compliance overall, in Canada, we're still seeing an increase in injuries. Some statistics that are published by workers compensation boards in Canada tell us a story, even as far back as 2009 when there were over 939 workplace-related fatalities across Canada.

In 2012, the last year that we have statistics on, that's grown to 977, so almost a thousand workers have been killed as a result of either exposure to chronic or acute occupational contaminants or due to a traumatic workplace incident. Those lives lost, paint the real picture and it's extremely unfortunate.

When we see there's a trend in increasing behaviour or frequencies in injuries, that's cause for alarms for sure.

Why Shortcuts in Workplace Safety Always Backfire...

Susan: Interesting that you say compliance is up but then so too are the fatalities. In your opinion, what's the root cause of this? I mean, almost a thousand individuals die each year in workplace accidents in Canada. Why is this?

Doug: It's an ongoing problem and a situation that requires our immediate attention; obviously there are families and workplaces that are being impacted, irreversibly, by every single incident. In the construction industry, specifically in Ontario, where I'm from, the market's "allowable lost time injuries" are also on the rise over the past couple of years.

As companies become more aware of their legal responsibilities for injury reporting and enforcement by the Ministry of Labor in Ontario continues to escalate, the expected trend in reported and allowable lost time injuries will likely continue to increase.

It's extremely competitive here in Ontario in the construction markets specifically. There's been a long-standing habit of underreporting when it comes to workplace injuries. Everyone knows this is against the rules. It's classified as Workers' Comp Fraud and it continues to this day.

A lot of employers are becoming more aware of this and are battling it on the front lines more effectively. Business owners are, more often now, forced to consider more effective ways to control unsafe workplace behaviours and conditions and put a stop to reportable lost time injuries through prevention, instead of hiding the incidents.

There are many contributing factors to an increase in workplace injuries and even, as I stated earlier, workplace fatalities. One reason for the rise in recordable injuries is due to an increase in the demand on business services in many industries which translates to an increase in demand on employee productivity which leads to people rushing, and as a result injury frequencies go up. Another reason is the cuts in company resources and expenses. Everybody's tightening their belts, Susan. And unfortunately, that means shortcuts which often lead to avoidable injuries.

Another reason for the increase in workplace injuries is employees are wearing many hats in various roles in their company. Businesses try to boost profits in the economic decline and they do this often at the expense of safety. The disconnect between production and safety is undeniable and continues to be the norm. Managers are also taking on more responsibilities in their jobs and as a result there's a large gap in the qualifications and competency when it comes to managing workers and performing job tasks.

And lastly, there's an aging workforce here in Canada and complacency can play a big role in injury frequencies. The 50+ demographic definitely has an impact on injury statistics, all you need to do is look at the numbers.

Susan: You would think with that as technology continues to improve, workplace injuries would decrease. Is that not the case Doug?

Doug: I think the statistics show the opposite. We're not able to effectively curtail the frequencies, and the likelihood of these injuries is happening not because of just those factors I mentioned earlier. When you dig deep you quickly realize adequate training and worker instruction is really lax.

Scheduling safety training becomes a decision made more often to get employees back to work, not to improve worker confidence, performance or safety unfortunately.

Many employers here in Ontario are more or less buying workplace safety training cards. They're not putting a lot of effort into designing the content or prescribing a set curriculum that reflects real workplace specific hazards.

Also, there's far too wide a range of available safety training standards. For example, with Fall Arrest Training for people who work at heights, you can find someone to instruct your workers from anywhere between 2 hours and 16 hours, depending on how much you want to spend and what curriculum you want to cover. Price shopping is still the leading factor for many employers when choosing safety training.

As long as there are training providers willing to cover any amount of safety training content in an 8 hour classroom session, there will continue to be employers delivering inadequate training. I keep seeing evidence of these trends when meeting client prospects on a regular basis. The training documentation I review with management often shows past training courses covering topics like WHMIS, Working at Heights, TDG and Propane Handling all in an 8 hour class for 15-25 workers at a time. It's hard to imagine these workers leave the day with anything more than a training wallet card in their pocket. Effective comprehension and learned behaviour cannot be taught in this type of setting.

The safety training available in many industries is far too varied, in my opinion. The majority of safety training being delivered to Ontario workers is far too generic. It's not designed with workplace specific hazards in mind. Not to mention there is often little follow up to validate competency and not enough employee instruction on safety equipment or emergency response. This has a huge impact on worker confidence and ability to execute their jobs with safety in mind and does not reinforce safe behaviours.

We are only this year, beginning to see industry developed safety training standards for topics like Working at Heights. Even still, there is a long way to go, and employers need to take the lead, not the back seat in developing relevant hazard controls and defensible safety training.

Why Don't Companies Embrace Workplace Safety?

Susan: Interesting. Why would a company *not* invest in workplace safety? Seems like this should be Smart Business 101, but from what you are saying that's not what you are seeing in the field.

Doug: There's no doubt about it. I've been a consultant in Ontario workplace safety working with various industries for over 15 years and the same reasons that existed 10 years ago are still apparent today. Companies have a misconception, even fear that compliance will be cost prohibitive or they will spend money on safety and get no return, so they see it as an overhead expense and a direct loss. It's a misconception shared by a lot of employers who just don't understand the basic principles of strategic business planning, forming habits or human psychology.

There's a fear that they won't be able to effectively compete and actually the opposite is true. Based on my experience, companies who strive for excellence in safety enjoy more loyal employees, they enjoy a more efficient and productive workforce. Their competitive ability is improved not reduced by well thought out safety investments.

A lot of employers lack an understanding of basic workplace safety principles required to lead change within their organization. So even if they want to adopt a proactive workplace safety culture, they often lack the leadership skills to implement a safety campaign successfully.

Unless you've been trained or have experience in positive safety culture and leadership, it's an unchartered territory for many senior executives in a variety of industries.

Susan: Right. You're saying these companies, they specialize in what their unique ability is, whether it's manufacturing or construction, they don't necessarily have a fundamental understanding on how to incorporate safety into what they already do.

Doug: It's becoming more and more common now as well to acknowledge the value in mentorship. Mentorship is a word that I'm hearing a lot lately especially in the economic growth organizations who are trying to support diversity and deliver sustainability to a lot of local economies and safety is no different.

More now than ever, people who are successful and are growing their businesses and brands, who are true visionaries, have mentors. If you're lacking in a skill set, whether it's accounting, marketing or branding, it's proven that if you find someone who can help you fill that gap and coach you through unfamiliar territory; you're going to be more successful at delivering goals and hitting your milestones.

A lot of business owners think keeping the workers safe is all up to them as well. This can cause anxiety in managers and often they see doing things safely as a burden. They don't understand how to fully engage their workforce in a culture shift. The old school, "Do as I say" approach to leadership is still the predominant mindset in some industries and it's a hard habit to break. That macho and authoritative figure who has been seen as an unapproachable CEO or boss in the past does not help drive positive change, but that is what many Canadian employees know and still see, especially in the trades.

Styles of leadership that are becoming more understood and sought after are ones like the "R+ Style of Leadership" brought about by companies like Bill Simms in the US; they helped evolve positive reinforcement and incentive programs that reward positive behaviours.

Focus on Sustainable Safety,
Not Compliance...

Susan: Doug, let me ask you this. You mentioned earlier that a lot of companies "are buying their compliance programs." What does that mean exactly? What should they be focusing on?

Doug: Good question, Susan. The track record that I see as a third party observer in a lot of circumstances is employers get so caught up in compliance because it's easier to measure. You can audit your safety programs compliance nowadays simply by filling out a questionnaire.

The agencies like Workplace Safety Insurance Board and the government associations that help support safety improvements provide all kinds of checklist and audit tools. In my opinion, that's taking too narrow a focus. You have to really look at safety from a business perspective and sustainability is what's going to help grow companies successfully and maintain their competitiveness. If you focus on compliance first, it's harder to integrate safety and productivity.

Understanding behaviours and perceptions requires employers to observe and engage their workers in dialogue and ask for their feedback. That's not as common as you might think in many organizations. Businesses who are investing in safety just to pass an audit or get Training Wallet Cards are missing the point which, in my opinion, is injury prevention. It's saving lives. It's getting a handle on unsafe behaviours and conditions in the workplace rather than measuring whether your program is compliant. That's a really key realization if you want to invest in safety for the right reasons and see a positive return on your investment.

You can put your workers through online training for Confined Space Entry, as an example, but they're not necessarily going to learn applicable procedures that reflect your specific workplace or even get information on the appropriate emergency response plan that's mandatory and critical to save lives in those dangerous environments.

The online courses can't teach workers how to use a self-contained breathing apparatus, for example; that takes hands-on experience and qualified and competent trainers as well as regular practice and practical mock scenarios. I think I can speak for a lot of consultants out there, online training is still a sought-after solution to fill the training gaps in many different industries and it's unfortunate and disappointing to see that employers still treat online courses (as well as the "self-teach" video & workbook combos) as a complete solution for filling their training needs and passing a compliance audit.

I believe, when your employees' lives are on the line daily and they often are, especially in the construction and industrial service industry, an online course isn't enough. Let's be real. That can be a dangerous path sometimes and often gives workers a false sense of security because they are told that "training" is in place, so get back to work.

Susan: It's almost like these Wallet Training Cards provide the organization with a false sense of confidence in their safe work habits. They think they're being compliant but they're barely scraping the surface, aren't they, if that's their approach?

Doug: Absolutely. Training Cards are becoming a commodity. They're not meaning anything more than you pass the minimum checklist. That's not making any workplace safer.

Workplace Safety Leads to Higher Profits and Increased Morale...

Susan: Interesting. The companies that invest in workplace safety actually see a higher return than those that just play lip service to safety. Can you say a little bit more about why that is?

Doug: It's becoming a fairly well-understood phenomenon especially with top 100 employers. The brand names that know they can no longer afford negative publicity decide to invest in safety for the right reasons and reap immeasurable rewards. There are several examples of relevant situations where safety can directly affect profit in a positive way.

A lot of large employers are pre-qualifying contractors nowadays. They take a short list of all their potential subcontractors and they're asking for evidence of workplace safety be present (and audited in many cases) before they even accept bids or tenders. That's something that's becoming very mainstream here in Ontario, especially over the past 5 years.

Bottom line is many employers are losing bid opportunities they used to win simply on price. They're losing high-level projects where there's lots of margin because they can't pass a simple compliance audit. The pre-qualification process plays into an organization's profits in a direct way.

Susan: If you aren't even allowed to bid on a project that in the past you would have been able to, that will have a pretty big impact on your bottom line.

Doug: Exactly. Safety is definitely one topic that I think no one will argue when done properly, will have a positive impact on efficiency in the workplace. Employees feel less like numbers and more like they are part of the team. When workers voices are heard and their opinions are listened to; and they see that their contribution to workplace safety is having a positive impact on progress they feel valued, motivated and their performance goes up. Whether it's joining the Health Safety Committee or participating in a workplace inspection, hazard observation, hazard analysis etc. being engaged fosters collaborative decisions and good habits. This scenario is one that can shift business culture and help growth and sustainability in so many different ways.

Other scenarios that definitely can impact the company's bottom line when safety is managed properly include the morale factor. Companies that invest in safety see an increase in employee morale. Decisions are made with people and production in mind first and foremost.

And from a much broader perspective another way Workplace Safety programs affect a positive return is when a business's reputation is at stake. A lot of the large organizations that excel at safety excellence receive a certain level of brand recognition and admiration from their peers and when competing for high-profile projects. Not to mention the potential benefits of relationships with safety-minded general contractors here in Ontario that measure safety performance throughout the project. You can actually be thrown off these high profile jobs if you are not performing to a recognized standard and you're given deficiencies on a regular basis. Bad press is bad for business. Sometimes you only have one chance at a first impression especially when you are a small business, trying to play with the bigger guys.

Susan: Sounds like companies that pay lip service to workplace safety may not even be in the running on a lot of jobs.

Doug: That's right Susan. It used to be common place in order to get involved in high profile projects all you had to do was get your CEO to sign an acknowledgment that says we work within the rules. That's all you had to do.

That commitment on paper now is no longer enough. I would say for the majority of the high profile jobs in the construction market here in Ontario, you have to submit your whole program to a fairly detailed and intense audit process.

Susan: As it should be.

Doug: They're looking for evidence of an Internal Responsibility System. You have to prove the things you say you are doing are reality.

Workplace Safety Is the True Measure of an Organization's Success...

Susan: I think it's interesting that they think workplace safety is going to get in the way of efficiency and profits but, as you pointed out, how efficient are they going to be when morale is low or if they have a workplace injury? That's when the real inefficiency comes in, right?

Doug: Absolutely. There's a little story I'll share about an American company that saw some huge change back in 1987. I'll paint the stage for you just briefly.

Well known aluminum manufacturer, Alcoa in the US, had an investors meeting in '87 at a posh Manhattan hotel where they were introducing their new CEO. They were shocked, the investors and the stockholders who were present, to hear Paul O'Neill, their new CEO say,"If you want to understand how Alcoa is doing, you need to look at our workplace safety figures."

His statement, "If we bring our injury rates down, it won't be because of cheerleading or the nonsense you sometimes hear from other CEO's. It's because the individuals at this company, in Alcoa, have agreed to become a part of something important. They've devoted themselves to creating a habit of excellence and safety will be the indicator that Alcoa is making progress throughout that shift."

Paul O'Neill saw workplace safety being the true measurement of their performance and how they'd be judged down the road.

Susan: Very good.

Doug: In that statement, in that meeting, on that day in Manhattan, he never promised that his focus on workplace safety would increase Alcoa's profits, but five years in, they saw profits and their shares value increased by over five times.

Four Common Workplace Safety Misconceptions...

Susan: Very nice example of a company who embraced safety and didn't scorn it and as a result is much better for it. Since you've been in this industry for so long Doug, can you share with us some misconceptions companies may have about workplace safety?

Doug: Absolutely. One of the biggest misconceptions is "Safety is Job One" and it's the primary consideration of an organization that's successful. I call that the *Bumper Sticker Mentality*. You drive down the highway anywhere in North America and you see the trucks of brand name companies with the bumper stickers: *Safety is Job One, Safety 1st, Safety's Priority One at this company*, and so on but a lot of the businesses who proclaim that Safety is Job One, when you dig down and look at evidence, are not doing what they say.

The statement "*Safety is Job One*".., although it's a great statement and a feel-good measure, I think it's impractical in almost every industry to say that it is the most important thing. Profitability and a business's sustainability have to remain the priority of the CEO or stakeholders and shareholders, because without the business' sustainability, there's nobody to protect. That's one example of a big misconception in the industry in my opinion.

Another would be; Safety must be developed and driven from the top down to be effective. Although I do believe that you need strong leadership and the decision and commitment to safety have to be practiced by executives and senior management. Effective changes, the changes that last in my opinion, are born from the ideas and participation of the workers.

The principle of listening has to be part of what managers and the supervisors do every day while executing their managerial responsibilities. If they're not collaborating with workers on solving problems, they're never going to gain trust. They're never going to put in place relevant solutions.

A great example would be when you are changing a policy about what to wear on a job site. If you weren't wearing safety glasses on the jobs in the past and management decides that safety glasses are going to have an impact on injury frequency, simply enforcing that policy and saying this is going to change as of today because we say so..., you're not going to get a lot of buy-in. Effective change just doesn't happen from a memo.

What a lot of companies struggle with is that changes or improvements by management aren't seen at face value by the workers. I believe that if you were to consult workers before making change in policy and get their buy-in and feedback, the decision would be a collaborative one and consensus would definitely then support your ability to enforce change.

If you get workers to share their thoughts and their feedback about change, they're more likely to allow that change and to support that change for the long haul.

Other myths include; when competing for new jobs and projects that things will be more difficult if we, as a company, spend time and money on safety. I believe that it actually becomes easier. If you have an added sense of pride and confidence, it eliminates the anxiety of being scrutinized, that's an advantage, and competition is something that makes successful companies want to reach goals and want to establish change.

Lastly, another myth would be that lack of injuries equates to a safe workplace. That's just not true. Many companies have killed or injured their workers without any prior injury experience or prior history. In most cases, in fact, there exists in those scenarios poor or non-existent safety culture or underreporting and very little due diligence or defensible training. Those are four more very common myths and misconceptions about workplace safety.

Susan: Right and I can see where a company that didn't have a long history of workplace injuries could make the mistake of then thinking their workplace was safe. That's not always the case.

Doug: Absolutely.

Susan: You actually are hitting at the real solution here, which is it needs to be a fully-integrated solution. Safety can't just be a bumper sticker slogan; it needs, as you pointed out, to include the workers. You can't have a solid workplace safety campaign without collaboration and commitment from all levels.

How the Real Test of Your Safety Program Isn't Zero Injuries but Rather: Zero Unsafe Behaviours and Conditions...

Doug: Right. One of the common goals you'll hear leading companies adopt is zero injuries. When in fact, if they put an effort in achieving zero unsafe behaviours and conditions, those zero injury goals would be a result naturally of controlling behaviours and conditions.

I like to get companies to not commit to zero injuries but commit to change. Commit to measuring and monitoring performance on a different level.

Susan: Can you say that again, Doug? I think that was really key. You said that they shouldn't use Zero injuries as the measure, they should use...

Doug: Zero unsafe behaviours and conditions.

Susan: Zero unsafe behaviours and conditions. If a company could get to a place where zero unsafe behaviours and conditions existed, that's really where the change happens, right?

Doug: That's right. Focusing on zero injuries is a reactive response. To be proactive in injury prevention, you have to start identifying and fixing what you observe is wrong on a daily basis, not once there is an incident.

Here's What a Workplace Safety Campaign Consists of...

Susan: Very interesting. Can we talk about what a workplace safety campaign consist of? I assume it's pretty comprehensive but maybe you could break it down for us?

Doug: A workplace safety improvement campaign, if you want to call it that, is something that differs for every single business. It must be customized in order to be relevant and effective.

The process of making improvements that are sustainable definitely needs to also reflect the company's individual capabilities. A very important part of that is understanding the perceptions workers have about their workplace.

Administering something like a perception survey helps paint a picture of where you need to start change. That's a difficult decision for a lot of employers. They're scared, in fact, to hear what their employees really think. The perception survey must also be completely anonymous and the setup for that first step has to be one of trust and support because if you simply sent out a perception survey to workers and they don't understand why it's being done, well, you're going to have a lot of people put up some defences and not want to answer truthfully.

A proper workplace safety campaign requires an understanding of where things are today in the organization. The next thing you've got to understand is who do you want to be known as in your industry? What kind of reputation are you seeking, what legacy do you want to leave behind? How much do you want to change within your business and it's got to start, in my opinion, with the culture of your company because if you invest in safety for the right reasons, your culture will shift and become one that sustains change and fosters true collaboration.

The process of making a real impact on sustainable change is one that involves learning a bit about yourself, learning a bit about your company and the people whom you have as stakeholders. It involves setting a plan in motion that is created through a collaborative process. It's one that can't be designed solely from the top down.

You're leadership team can have and share ideas and you must show and even demonstrate leadership at the top level, but the process of making change has to be one that involves all workers. The process of developing those changes has to be one that incorporates communicated (and encouraged) workplace specific concerns, identified workplace specific hazards and relevant and effective workplace safety controls.

Take away the possibility of your safety control solutions being generic and you will see solutions that are 100% relevant and effective. That's unfortunately where a lot of employers fall short, they take an industry best practice or they buy into training they see their competitor's delivering and just because they see it's available and convenient, they go buy it themselves without thinking about the relevance and whether it's applicable or appropriate for their business.

Susan: They want to do just enough to not get in trouble.

Doug: Right there, that's a real key point to understand and acknowledge, why you are going to make the commitment in your company to invest in the safety in the first place. It's got to be because you're trying to protect the lives of your workers. It's got to be because you put a face to their name. It's got to be because you want their trust. With trust comes loyalty and performance, I guarantee it.

A shift is safety culture can't happen and be successful if you're doing it because you're avoiding jail time or you're avoiding fines or you want to pass an audit to get a job. The afore mentioned motivators, if embraced, are what sustain a business' future growth.

Susan: Interesting. Do you get a lot of resistance when you share this concept with potential clients?

Doug: Absolutely at first. It's the first thing many employers say - "We can't afford the time, money and commitment involved in making change impactful" but once they understand the process, it makes 100% sense and all fears quickly disappear.

If you go to your workers and say, "Hey, guys. It's time for us to shift our priorities and put a new focus on workplace injury prevention and we want you guys to be the leaders in this," you're going to have a lot more employees want to become engaged. They're going to want to share their opinion and what you'll find is there'll be better, stronger relationships between workers and their managers. It's a culture shift.

Susan: Very interesting, Doug. Much more effective than trying to use safety as a slogan with no real meaning behind it. For the guy that comes to you and says, "Oh gosh, you know what? I've just been tasked to get this handled by my boss; can I just get a quick quote?" What would you like to say to them? You want them to take a much broader approach to safety and I think they need to hear it.

Doug: I like to say to anyone who puts up obstacles, consider this. If that person/prospect has children and those children start working in a part-time workplace or even a full-time summer job, would you want their employer to make safety something they invest in just to remain compliant or would you rather they work for someone who has a culture of safety that's built on trust and sustainability? I think it's a no-brainer.

Workplace Safety Works For Every Company in Every Industry...

Susan: It is kind of a revolution what you're suggesting here. You're trying to shake up the companies who are approaching this safety thing as just something to meet compliance. You want to turn that around and say, yes, you'll meet the compliance but you can do much better than that, your employees deserve much better than that, and our families deserve much better than that. You can actually make lasting change that'll affect your bottom line if you come at it from the right approach and use a collaborative team and include workers in the process.

Doug: One hundred and fifty percent, Susan. These changes are available to every single company in every single industry of every possible size and these changes are opportunities to make a lasting impression and to strengthen your business' sustainability.

There's no doubt that image plays an important role in influencing perceptions, and if you can promote a positive image to both your workforce, to your competitors and to your community, you're going to leave a legacy that's sought after and well deserved.

Susan: I'm very glad you're highlighting this for us Doug. This isn't just for those in an industry that necessarily has high risks.

Every organization can benefit from looking at their safety program and really making it part of their culture versus something they visit once a year, update a manual and then off it goes to the drawer to wait until next year.

Doug: Right. Absolutely. Compliance is important but it is far from the only element involved in a sustainable safety program. It's important but it certainly isn't or shouldn't be the driving force.

The driving force is trust, it's collaboration, it's culture and, honestly, inescapable commitment from the company as a whole. Bottom line for me as a coach, a husband and a father is this; every Canadian..., every human being, has the right to a safe place to work. This is embedded in the World Health Organization's fundamental principles and is a top priority in their leading international strategies, why some Canadian employers are still not accepting it as best practice I will never understand.

Here's Exactly How to Get Your Workplace Safety Program Going...

Susan: Very well said Doug. Can you share with us how your safety program works?

Doug: It's a pretty simple process, really. Although it's customized for every single company and each employer's needs, there are three main components.

First, they go to https://podio.com/webforms/6303769/491281 and complete a confidential safety survey that highlights gaps that exists in their current system. The results of the survey will provide us with the baseline on where to work from, where they're starting from.

The next step is to schedule a 90-minute meeting with me or a member of our team where we outline and develop a workplace safety program strategy that's tailored to the client based on the audit results. It's a collaborative process and we involve key management personnel and at least one representative from your workforce in the priority setting process.

The third step is we develop a project team to bring your business into compliance with whatever workplace safety standard is applicable to your industry. We collect the evidence, we start documenting your health and safety program deliverables and this is where a lot of employers without coaching fall short, either because they don't have the confidence or the tools to make those changes effective and actually measure the progress. We welcome all input and put the plan in writing. From day one of your Workplace Safety Revolution you have a formal plan online that evolves daily and documents milestones, its shows real change is happening and who is involved in the program deliverables.

If you're going to develop a program that is evidence-based, you've got to have the tools and you have to get all levels of workers involved throughout the process.

Susan: Agreed. Any final comment Doug?

Doug: Susan, I think where I'll leave it is take the next step and create an opportunity for change in your organization. It doesn't necessarily have to be with Workplace Safety Revolution Inc. but it should be for and with your employees because every worker deserves a safe place to work and they will appreciate the investment for the right reason and so will their families.

Susan: Thank you, Doug. This has been very eye-opening, I hope everyone takes it to heart, it should be mandatory that every organization takes a look at their safety program at this deep level rather than just doing the lip service. It scares me that there are organizations that may be skimping in this area.

Doug: You know what? I see a lot of consultants out there using scare tactics to get employers to invest in safety and that's also an ineffective and unsustainable approach. There has to be 100% transparency but there also has to be 100% positivity. My strategies and tools are designed to be stepping stones to business excellence; they are not mandated solutions or designed as the "only way" to achieve safety compliance. I'm just trying to be a coach to the coachable. The changes I suggest to my clients need to be looked at as opportunities for improvement, nothing more. Each client's progress depends much more on the businesses ability to demonstrate commitment, collaboration and culture.

Susan: Right, and if they do it right, it'll impact their bottom line in a very positive way. I like it. How can they get in touch with you if they have more questions?

Doug: They can visit my website http://www.workplacesafetyrevolution.com. Or they can also contact me through email at doug@joinwsr.com.

You'll find me on social media for sure, everywhere from Twitter, to LinkedIn, to Facebook. Google "Workplace Safety Revolution" and you'll have no problem tracking us down.

Susan: Thanks again Doug.

Doug: Thank you, I appreciate the opportunity.

Here's How to Take Control of Your Safety Responsibilities and Get Your Workplace Safety Program Compliant and Documented...

You already know your workplace safety program policies and procedures need to be updated. Your clients may soon start demanding to see your safety program evidence and training documentation before they will even look at your bid. You know you need to take a more critical approach to holding workers accountable but never seem to get around to it. There always seems to be something more pressing on the to-do list.

That's where we come in. We help businesses just like yours bring their workplace safety into compliance and put the tools in place to support not only sustainability but also continuous improvement and ensure development of an effective Internal Responsibility System.

Step 1: You complete a confidential safety survey to highlight any gaps in your current Occupational Health and Safety Management System and provide us a baseline to work from.

Step 2: We meet for 90 Minutes to outline and develop a tailored Workplace Safety Program Improvement Strategy for your organization.

Step 3: We work with your team to create a Safety Culture that will bring your business into compliance with the Occupational Health & Safety Act and document your Health & Safety Program progress. Responsibilities are delegated and we fill the prioritized gaps in your OHSMS.

Now you can begin working with confidence, because your safety program has you and your workers adequately protected and your business can start reaping the benefits, with just one call.

Join now and ***Lead... Change.***

If you'd like help, just send an email to: info@joinwsr.com and we'll take it from there.